AUSTRALIAN
Tropical
Rainforest Life

by
Clifford & Dawn Frith

Published in Australia by Tropical Australia Graphics

National Library of Australia Card Number ISBN O 9589942 2 6

Produced in Australia by Tropical Australia Graphics,
'Prionodura', Paluma via Townsville, Queensland 4816

First published 1983
Revised edition 1986
Third printing 1987

PREFACE

Having first met one another on the utterly unspoilt tropical coral island of Aldabra Atoll in the mid Indian Ocean, and subsequently travelled the world extensively including four years living in South East Asia, we were delighted to find the rainforested areas of tropical North Queensland as spectacular and zoologically exciting as anywhere we had experienced. We have now lived at Paluma, southern limit of many Australian tropical rainforest endemic plants and animals, for nearly eight years; here we study rainforest insects, plants, bowerbirds and other birds. For the first three years here Clifford worked as a post-graduate student of Monash University, Victoria, but is now a full-time freelance zoologist, artist and photographer. Dawn graduated from London University with an honours degree in zoology, followed by a Ph.D. degree in marine science. Our joint researches have included work on crustaceans, insects, amphibians, reptiles and birds; but birds now occupy us both fully about our rainforest home.

All photographs herein were taken on 35mm Kodachrome 64 (KR 135) slide film. Recent photographic work has been done with OLYMPUS OM 2n and OM-4 cameras and ZUIKO lenses from 35mm wide angle to 500 mm telephoto. Two Mecablitz 402 flash heads were used for birds at nests and larger stalked animals and birds; and two regular tiny hot-shoe type flashes were used for some frog, reptile, insect and other small subject work. All travel has been by Landrover.

Three animals we required for this book we were unable to photograph in the wild, and had to resort to using captive animals in the care of the Queensland National Parks and Wildlife Service, Townsville. These are the Herbert River Ringtail Possum and the Grey and spotted Cuscus; and we thank Hugh Lavery, Gavin Blackman and John Winter for their kind help in this regard. We are pleased to express our gratitude to Hans and Judy Beste, Bill and Wendy Cooper, Bruce and Neil Dingwall, Bruce Fuhrer, Betsy Jackes, David Mitchell, Andrew Taplin, and the good people of Ivy Cottage Tea House, Paluma for providing animals, information concerning their whereabouts, identification of subjects or valuable assistance and companionship in the field. Avril MacBain kindly proof read the text. We thank the C.S.I.R.O. Division of Entomology, Canberra for identifying some insects.

Since the publication of the highly successful first printing of this book we have also published a companion volume titled 'Australian Tropical Birds', and a small introduction to the environment book 'A Walk in the Rainforest' illustrated by David Stacey's line drawings.

Clifford and Dawn Frith
Paluma, Tropical North Queensland

Photo: David Stacey

1

Lowland rainforest, Daintree River area

INTRODUCTION

This portfolio represents the best photographs, of some subjects, taken during four years camera work in tropical North Queensland rainforest. The pictures are merely a small selection from a vast array of life that no modest book of this kind can do justice to. This is not an in-depth account of rainforest life and the text is not greatly ecologically inclined, but is presented solely in the hope of stimulating an appreciation of, and better still interest in, the subject. We hope the brief text answers most commonly asked questions about the pictures. Those wishing to learn more should consult books listed in the bibliography on page 66.

To the casual visitor, Australian tropical rainforests appear relatively lifeless; a sadly common misconception. For those prepared to sit and wait or gain a little insight before seeking it out, however, the breath-taking array of life only touched upon in these pages will reveal itself. This book is most certainly written with the absolute conservation of remaining Australian tropical rainforests foremost in our minds as an urgent requirement. Only expressed public opinion will save our rainforests, and public opinion will only be generated through appreciation and understanding. Hence this book.

Any sensible person can appreciate Australia's need to retain all remaining tropical rainforest, especially the few last virgin areas. This natural heritage is in the care of the Queensland Government and in this respect that State holds an immense responsibility of national and international significance to the peoples of the present and future generations. Contemporary forestry activity in tropical rainforest is not labour intensive, providing only few jobs; and producing very predominantly elitist commodities for people who can probably no more tell the difference between a genuine rainforest timber piece of furniture and an artificial or veneered one such as most folks today are content with. A priceless natural heritage is being destroyed in very large part for decadent usage.

Areas of tropical rainforest extensively clear felled are often lost forever, as all topsoil is subsequently lost by erosion and no plant life remains as a recolonizing stock. Forest severely damaged by selective felling may take several hundred years to regenerate to its original undisturbed conditions.

Australia is one of few politically stable and physically safe countries where people can enjoy tropical rainforest environments. Conceivably it will one day be the only remaining country with such readily available tropical forest attractions of world-wide tourism significance.

Coastal lowland rainforest, Cape Tribulation

Upland rainforest, Paluma

ly the grossly ignorant, naive or
lish can be content to think we can
in relatively small 'sample' areas of
diverse and luxuriant habitat for
ure tourist exploitation, scientific
deavour, educational or other
rposes. Rainforest ecosystems simply
not persist long-term under such
nditions. Many factors too complex to
cuss here, such as gene pools, gene
w, and local climate maintenance,
mand that most extensive areas of
nforest be left intact. Putting aside
ological, economic or aesthetic
guments, Australians and other
oples surely wish to ensure their future
nerations have the opportunity to
oose, to enjoy, experience,
derstand or study the tropical
nforests as a birthright.

Tropical rainforest is the richest of
earth's environments in plant and
mal life, together co-existing in a
atively stable complex ecosystem. It
vers some 13 million square
ometres of earth's tropical belt in
uth America, Africa, Indo-Malaysia,
onesia and Australasia. Past sea level
anges exposed land bridges between
stralia and New Guinea allowing floral
d faunal interchange. Many plants and
imals of Australian tropical
nforests, particularly those of Cape
rk Peninsula, occur in or are more
sely related to those in New Guinea
ests; the two areas collectively
nstituting the Australasian tropical
nforests. It should be stressed,
wever, that our tropical rainforests are
many ways unique with respect to
merous plants and animals.

Some 600 to 800 tree species occur
our tropical rainforests. The crowns of
gest forest trees interlock and form a
osed-canopy' that shadows the middle
nopy, subcanopy and forest floor
low. The upper canopy supports an
rial garden of climbing vines,
iphytes such as giant elkhorns,
ghorns and bird's nest ferns
plenium nidus (page 5), orchids,
osses and lichens. Woody climbing
es, the lianas, entwine their way up
o and then across the upper canopy
owing to enormous lengths.

Beneath the upper forest canopy
ht is greatly reduced, temperatures
e lower and humidity much higher. In
ver strata of the forest and on the
est floor innumerable other plant
ms compete for space, light and
trients. These include the smaller tree
ecies, saplings of potential forest
ints, lawyer canes, bananas, gingers,
ound orchids, various epiphytes (see
ge 6) including ferns and orchids,
osses and lichens, fungi and
merous other plants. Many climbing
ints festoon tree trunks by using
ndrils, hooks, or stem roots to maintain
ooting as they clamber upwards.

Bird's Nest Fern

Australian tropical rainforests occur along narrow coastal plains, foothills,
mountain ridges and tablelands from sea level to 1 600 (average 800 to 900)
metres. The now rare lowland coastal rainforests such as that at Cape
Tribulation extend to the sea edge, where they may even abut the salt-tolerant
mangrove forests (page 3). Lowland rainforests are characterized by robust
woody lianas, epiphytic ferns, the lovely fan palms, and strangler figs; and the
leaves of lowland forests are generally larger than those of the uplands, and
plants are more often deciduous. Some of these characteristics, notably the
fan palm **Licuala ramsayi**, can be seen in the picture of the Daintree area
forests on page 2.

Upland forests also characteristically have woody lianas and epiphytic
ferns; but tree ferns, climbing vines and mosses appear to be more abundant
(see pages 4 and 67).

BUTTRESSED TREE TRUNK

Buttresses are a feature of many trees in all tropical rainforests. These woody flanged extensions radiate outwards from the lower part of the tree trunk and, like the one pictured here, may reach large proportions; sometimes up to 10 metres in height. Buttress evolution and function are somewhat of a mystery. Although buttresses to some extent support a tree's weight in shallow soils by taking up strains and stresses during windy conditions, it was recently demonstrated that just as many buttressed trees fell down in cyclones as those without!

Shallow root systems extend out horizontally on, or just below, the forest soil both from buttressed trees and those lacking them. It is these shallow roots that appear to be an important adaptation to various soil conditions. Many rainforest soils are shallow and easily become waterlogged, and, generally, there is a greater concentration of nutrients nearer the surface than at lower depths. In such soil conditions shallow roots are more efficient than would be the more familiar deep tap root systems.

Photo : Paluma

STRANGLER FIG
Ficus destruens

Of some 16 native fig species in Australia, 6 occur in tropical Queensland. Strangler figs, climbing plants and epiphytes (plants growing upon other plants but not deriving nourishment from them) are all forest plants that utilise trees for attachment and support. The mighty strangler fig actually kills its host tree. Animals including fruit-eating birds and bats feed on, and so disperse, fig fruits and subsequently deposit the seeds in tree crowns, often 35 metres or so above ground. By germinating so high up in the forest canopy the strangler avoids competition with the numerous tree seedlings that start their struggle for survival on the forest floor. The young fig develops long thin cable-like roots that grow down the host tree trunk to the forest floor where they can readily absorb a plentiful supply of nutrients and water. Now the fig tree flourishes. Roots become thicker and interlace their way tightly around the supporting tree trunk, and eventually the host dies leaving a totally independent strangler fig which may live for several hundred years (see also page 69).

Photo : Paluma

BIRD'S NEST or CROW'S NEST FERN
Asplenium nidus

NORTHERN ELKHORN
Platycerium hillii

A tropical rainforest scene would be incomplete without the massive epiphytic ferns that spread majestically out from trunks and branches. Although these epiphytic giants mainly occur in the canopy, some, like those pictured here, may grow on trunks nearer the ground.

Roots of all epiphytes, such as these ferns, derive moisture from rainwater as it trickles down the tree, and absorb nutrients from rotting vegetation trapped by the epiphytic structure itself or from crevices in the bark of the supporting tree. The basket-shaped growth-form of bird's nest ferns (above) and elkhorns (below) provide an excellent catchment area for such falling debris; so much so that the plant itself supports its own miniature 'forest floor' leaf litter, and associated life. Insects, other invertebrates, and frogs take up residence amongst the damp decaying matter and in turn prove tasty titbits for many rainforest birds. For example the Eastern Whipbird, although generally a forest floor forager, often uses this arboreal 'forest floor' for seeking food.

Photo   :   Iron Range

NATIVE MONSTERA
Rhaphidophora pinnata

More commonly recognised as a popular decorative indoor plant, this large-leaved climbing Native Monstera occurs both in highland and lowland rainforests. Monsteras are root climbers; that is apart from their normal soil roots their main stems develop long flexible cane-like roots which firmly attach them to a supporting tree. The leaves of young monsteras are entire but as the plant matures they become attractively divided. Many monsteras produce lovely large white hooded flowers which are followed by big fleshy fruits, some of which are edible when ripe and are considered a delicacy.

Photo   :   Iron Range

LAWYER CANE or WAIT-A-WHILE
Calamus motii

Most people think of palms as tall slender trees fringing tropical sandy beaches and rustling in the breezes! Many such typical palms do occur in our forests, for example the splendid fan palms that are so typical of our lowland forest (page 2) and the particularly elegant bangalow palms (lower left in above picture).

Several species of palms, including lawyer canes (right and centre in above picture), are very common in upland forests (see page 4) and have become climbers. Stems and leaf ribs of lawyer canes are armed with viciously sharp hooks (right) which certainly deter any browsing animal from eating them. Long wiry tendrils with strong recurved hooks branch out and upward from the main stem and function like grappling hooks as a climbing aid. These sharp hooks latch onto nearby vegetation and so provide support for the plant as it grows upwards toward the light.

The name Lawyer Cane was reputedly given by early Queensland settlers who supposedly imply that once hooked by this plant one is as entangled as being involved in the legal process. This palm is also nicknamed Wait-a-while because of the delaying effect it invariably has on rainforest travellers. Stripped of the hooks the fibrous stem becomes 'rattan' from which cane furniture is made.

Photo : Paluma Range road looking toward Hinchinbrook Island

CLIMBING PANDANUS
Freycinetia excelsa

Another group of plants that climbs in rainforest is the climbing pandanus vines, that are more familiar as spiral-fronded trees called screw pines of the lowlands. Like the Native Monstera, they are root climbers. Stem roots are small and numerous and closely hold the vine to a supporting tree. The main stem gives rise to many loosely hanging branches that may reach a metre in length (left). The male or female flowers develop at the end of a branch surrounded by bright orange leaf-like bracts (below), the brilliant colouration of which attracts birds such as honeyeaters and rifle birds to the nectar supply and so ensure pollination. Fruit-eating birds such as the bowerbirds, Victoria's Rifle Bird, and Brown Pigeon are commonly seen eating the succulent ripe red fruits.

Photos : Paluma

TREE FERN
Cyathea species

There are approximately 800 species of **Cyathea** on earth but only 11 of them, restricted to along the east coast, occur in Australia. Of these, 7 are found in north eastern Queensland. One tree fern **Cyathea felina** previously known only from New Guinea and Malaysia, was recorded in Australia for the first time as recently as 1977 in forests of the Cape York Peninsula. Tree ferns occur in lowland and upland forests and their habitats range from deep dark forested gullies to dry forest fringes and creek banks in more open situations. Like the tree fern shown here, many favour sunny situations such as beneath openings in the canopy foliage.

Although the fronds of the tree fern are long, those of some other kinds of ferns such as the King or Giant Fern **Angiopteris** are even longer, reaching lengths of 5 metres or more. This massive fern, one of the largest on earth, can be seen commonly fringing lowland creeks, but is also found in upland rainforest. Fossil records indicate that members of its family are of a very ancient lineage, flourishing in forests some 300 million years ago. Besides the massive tree ferns and king ferns there is an amazing diversity of smaller ferns, growing on fallen logs, tree trunks, creek banks or directly from the floor. Some, such as Maiden Hair and filmy ferns are very delicate while others resemble the more familiar bracken ferns.

Photo : Mount Hartley, south of Cooktown

WHEEL-OF-FIRE
Stenocarpus sinuatus

One of the most spectacular flowers of all rainforest trees is that of the Wheel-of-Fire tree (opposite, top). This, popular ornamental tree, reaches heights of 30 metres or so in tropical rainforests and the brilliant profusion of its flowers provides an outstanding splash of colour in the green canopy. The Wheel-of-Fire belongs to the family Proteaceae together with the banksias and grevilleas. One of the most characteristic trees of some tropical rainforest areas is the Northern Silky Oak, worthy of mention here because the timber of this close relative to the Wheel-of-Fire has a magnificent grain and is used for making furniture, more so in the past than at present. The male Golden Bowerbird infrequently uses the creamy white flowers of the Northern Silky Oak as bower decoration (see page 53).

Photo : Paluma

BUMPY SATINASH
Syzygium cormiflorum

Several tropical rainforest tree species exhibit a peculiar phenomenon called cauliflory, in which the flowers and fruits develop from the trunk or larger boughs of the tree. In so doing cauliflorous trees, especially the smaller trees that are unable to reach the light and space of the upper canopy, display their flowers to understorey pollinators such as moths and shade-loving butterflies.

Bumpy Satinash (opposite, bottom) is an excellent example of cauliflory. It is a tall emergent rainforest tree in both mountainous and lowland rainforests, reaching 30 metres in height. It is suitably named because of the bumpy nature of its trunk. Masses of delicate whitish flowers grow out of these 'bumps' almost down to ground level. During the daytime these nectar-rich blooms attract such birds as honeyeaters and lorikeets; whilst at night they may be visited by the delightful Queensland Long-tailed Pigmy Possum, a tiny nectar feeding marsupial. This photograph shows a very young Bumpy Satinash in flower, a tree little more than a sapling which lacks the bumpy trunk of older individuals.

Photo : Iron Range

ORCHIDS

Our tropical rainforests support a wonderful diversity of these magnificent and often bizarre flowers. There are some 35 000 orchid species in the world and not surprisingly they represent the largest family of flowering plants, nearly a 7th of all those on earth. In Australia they occur in a wide range of climates and habitats, but they are considerably more diverse in the wet tropical and subtropical regions of the east coast. Their flowers exhibit enormous variation in size, shape and colour within the group. In rainforests they mostly live as epiphytes atop emergent trees, rooted between tree forks among canopy foliage, or encrust trees or boulders along mountain creeks. Some, however, are terrestrial; growing from the forest soils in sunlit patches.

The flowers of many orchid species mimic the shape and colour of their specific pollinators, such as wasps, bees, moths or butterflies, in order to attract them to their nectar. The basic structure of all orchid flowers is similar, with 3 outer sepals and 3 inner petals. One petal is usually quite large and acts as a landing platform for its insect pollinator. Male and female reproductive parts are united into a single column within the flower. When an insect that is laden with pollen from another orchid enters the flower it brushes pollen against the female part and so fertilizes it; and on its departure it brushes against the male part and in so doing collects pollen to transfer to the next bloom. Unwittingly the pollinator perpetuates the orchid species.

The two species illustrated here both belong to one of the largest groups of Australian orchids, the genus **Dendrobium. Dendrobium bifalce** (above) was found flowering in July in lowland forest; and **Dendrobium adae** (below) is a highland species confined to North Queensland which we found flowering in August.

Photos : Iron Range & Mount Hartley peak

MISTLETOE

There are 1 300 or more mistletoes in the world but there are only about 60 of these in Australia. Although mistletoes carry out photosynthesis (see page 14) like all green plants, they are partial parasites in as much as they derive their water and nutrients from a host tree to its detriment. The illustrated rainforest mistletoe has particularly large flowers, is conspicuously colourful, and is quite unlike the more familiar dull green mistletoe we commonly associate with Christmas. The Mistletoebird, a tiny attractive species found throughout Australia in many situations including tropical rainforest, is particularly adapted to feeding on and dispersing mistletoe berries and with little doubt includes the fruits of this lovely species in its diet.

Photo   :   Paluma

STINGING SHRUB
Dendrocnide moroides

Stinging shrubs may be encountered in sunnier areas of tropical rainforest where trees have fallen or been felled, or along tracks and rainforest fringe areas. The leaves and stem of this shrub are covered by stiff hairs which, if touched, inflict a painful sting like an extremely severe stinging nettle. The plant manufactures its hairs from mineral silica, the chief constituent of glass. If one is unlucky enough to brush against these hairs, their tips penetrate the skin, break off, and an irritant poison is released. Even today there is no effective antidote although antihista-mines do help to some extent. Bushmen recommend rubbing the sticky gum of the cunjevoi plant onto the stung area and then peeling it off, thus hopefully pulling out some of the spines. Adhesive sticky plaster can be similarly used but these techniques do not successfully nullify the pain. The effect of the sting may last for weeks or even months and some people may react more severely, with swelling lymph glands, difficulty in breathing and sometimes even severe shock. Visitors to tropical rainforests should be aware of the possibility of stinging shrubs at all times.

The poisonous hairs do not affect all animals. Some spiders commonly walk over the hairy surface, two species of beetle and the caterpillars of the White Nymph Butterfly feed on the leaves and its raspberry-like fruits are eaten by the Spotted Catbird. The Green Possum (see page 61) also includes the leaves of the larger Giant Stinging Tree, **Dendrocnide photinophylla**, in its diet; obviously without any ill-effects.

Photo   :   Paluma

FUNGI

Unlike most plants, fungi lack the green pigment chlorophyll, and so are unable to carry out photosynthesis; the means by which green plants build up their food from carbon dioxide and water using energy from the sun's rays by a series of chemical reactions in the chlorophyll. Fungi feed as saprophytes or parasites; that is, they absorb food from dead and decaying plant and animal matter or directly from living plants and animals.

Each fungus consists of fine microscopic filaments called hyphae which together form a mass referred to as the mycelium. Mycelia spread out over and into the surface of a substrate such as dead leaves, logs, or diseased trees; not just by mechanical force but by secreting digestive enzymes which may start the decomposition process. Fungi therefore play a vital role in the rainforest ecosystem because by breaking up dead and decaying plant matter the all important nutrients are released back into the soil.

At certain times of the year, mostly during the wetter months, fungi produce fruit bodies and it is these we recognise as, and call, fungi. Each fruit body contains millions of microscopic spores, the equivalent of seeds in flowering plants, which when mature are dispersed by all kinds of mechanisms; be it by wind, rain-splash, vibrations, fruit-body explosion, or by insect or animal movements. Fruit bodies vary greatly in size, colour and shape. Although many forest fungi are shaped like the more familiar mushrooms (page 17, top) and toadstools, many more are quite bizarre resembling staghorns (page 14, top), organ pipe coral (page 14, bottom), slime (page 15), conical spikes (page 16, top), or shell like (page 17, bottom). Fruit bodies may be white, yellow, orange, red, brown, blue, purple or almost any other colour, and besides these different shapes and colours some species even luminesce at night! Very little is known of some of the more peculiar fungi, particularly those in tropical rainforests, and many of them have as yet not been named let alone studied to any degree.

Photos   :   Paluma

MAIDEN VEIL FUNGUS
Dictyophora indusiata

The Maiden Veil (right) is perhaps the most bizarre and photogenic of all rainforest, if not all, Australian fungi. Aptly call Maiden Veil or sometimes the Crinoline Fungus, it belongs to a group of fungi called stinkhorns, on account of the terrible smell they emit. These rarely seen fungi require a substrate with a high organic and moisture content. The white phallic-shaped fruit body, some 20 centimetres high, emerges from the ground during the night and spreads its lacy veil. Almost immediately, flies are attracted to the foetid smelling slime that exudes from the stinkhorn cap. Landing on the lacy veil, which itself probably acts as an attractant, flies walk upwards onto the cap where they feed on the foul slime. When the flies leave, the fungal spores stick to their bodies and are so dispersed. By the following mid-morning the veil collapses and the fruit body disintegrates.

Photo   :   Paluma

INSECTS

In 1970 the number of Australian insect species was considered to be 54 071, a mere fraction of the vast insect fauna of the world which constitutes about three-fourths of all animal species on earth. Tens of thousands of insect species may, thus, occur in our tropical rainforests. We can do no more here than present an example from a few of the better known groups.

DRAGONFLY
Neurothemis stigmatizanus

Dragonflies belong to a group of insects called Odonata and occur in a variety of habitats including rainforests throughout the world. In tropical rainforests they are mostly seen along creek beds, tracks and in clearings. These sunlight-loving, day flying insects are often seen darting from one grassy perch to the next, while others hover over vegetation, motionless, save for wing and head movements. Their bodies are long and slender, and their two pairs of semi-transparent wings, often brightly coloured red, blue, orange or brown, are well developed for powerful flight. Dragonflies feed by catching small insects, such as flies, on the wing and their strong jaws are well adapted for dealing with such animal foods. They rarely stray far from water, in which the females lay their eggs and the subsequent nymphs develop by predating other freshwater dwelling invertebrates and even small fish.

Photo : Iron Range

LONG-HORNED GRASSHOPPER
Phyllophora species

This Long-horned Grasshopper (bottom) belongs to a group of insects called tettigonids of which there are some 5 000 species on earth, and whose close relatives are the true grasshoppers, crickets and locusts. The bright green body and finely veined leaf-like wings of this species provide it with effective camouflage amongst leafy vegetation, on which the grasshopper also feeds.

Photo : Iron Range

STAG BEETLE
Phalocrognathus muelleri

This rare Stag Beetle belonging to the beetle family Lucanidae, is restricted to North Queensland rainforests, where it ves among rotting logs and tree stumps and is eagerly hunted by collectors. Its large size and splendid metallic green and urple body sheen make it one of the most attractive beetles in Australia. The strong curved mandibles, or jaws, are notably onger in the male than the female. Despite this armoury, however, stag beetles are predominantly vegetarians. Most lucanids re nocturnal and this particular male was attracted to our house lamp on the forest edge.

Photo : Paluma

LEAF BEETLE
Stethomela species

This mating pair of iridescent leaf beetles, belonging to the beetle family Chrysomelidae, is just one example of numerous species of beetles of various families found in our rainforests. Beetles are found, both as adults and larvae, among understorey and canopy vegetation; in flowers, mosses and ferns; beneath bark; and within rotting vegetation, living trees, leaf litter and soil. One group of beetles is solely confined to large bracket-shaped fungi that occur on vertical tree trunks or fallen tree logs.

Photo : Paluma

PAPER WASP
Polistes tepidus

The nest of this paper wasp species is formed of a single horizontal comb of hexagonal shaped paper-like cells attached to the support above by a short stalk (below). In this picture worker members of the social colony line the outer edge of the comb in a defence posture to guard their nest against would-be predators, in this case the photographers. The milky white eggs, which can be seen suspended within the cells, develop into paper wasp larvae. Workers feed these grubs on chewed up caterpillars of other insects and, when fully fed, the larvae spin a silken cocoon about themselves in which they develop into adult colony members.

Photo : Iron Range

HERCULES MOTH
Coscinocera hercules

This newly hatched male Hercules Moth (left), belongs to the moth family Saturniidae and is found only in far North Queensland and New Guinea. It is one of the largest moths in the world and is the largest Australian species. Still perched on the cocoon from which he has just emerged, the moth pumps fluid into the limp wings until they extend and become hard and dry. Male Hercules are slightly smaller than females which have a wing span of about 25 centimetres, and differ from them by having much longer feather-like antenna and more tapering tail-like hind wings.

Females secrete chemicals to attract males which detect the females' scent with their long antennae. After mating a female lays some 80 to 100 eggs on the leaves or stems of 6 to 8 rainforest tree species, which are the caterpillars' only food plants. Sadly, females die after mating and egg laying as they have no mouth parts and so are unable to feed. They can only live as long as their fat deposits last. The blue-green caterpillars grow to some 10 centimetres long before they spin cocoons.

Photo : Paluma

FRUIT-SUCKING MOTH
Othreis jordani

This attractive insect belongs to the moth family Noctuidae and is just one example of the bewildering variety of moths that live in the tropical rainforests of North Queensland. The Fruit-sucking Moth pictured here is sucking up juices from an over-ripe fruit of the Native Banana. Closely resembling the cultivated kind, Native Banana grows wild in lowland forests, especially in damp swampy situations. Wild banana fruits are small and thin and contain numerous hard seeds which make them scarcely edible to humans. It is, however, a favourite food of many rainforest fruit-eaters including these moths, the long strong sucking tubes of which are well adapted to penetrate through the skins of such rainforest fruits.

At rest the green and brown patterned leaf-like upper wings of this moth are folded over the back of the body and so conceal it from predators. When disturbed, however, the moth flashes open its upper wings to reveal the brilliantly coloured hind wings in order to frighten off predators.

Photo : Iron Range

CAIRNS BIRDWING BUTTERFLY
Ornithoptera priamus

This birdwing butterfly of the family Papilionidae is one of the largest and most spectacular of all Australian butterflies, of which there are nearly 400 species. There are only 2 birdwing species, of the predominantly New Guinea and South East Asian group, in Australia and both are fully protected by law.

The wing span of the female Cairns Birdwing Butterfly reaches about 20 centimetres across from tip to tip and the male, pictured here (below), although slightly smaller is far more colourful. Forewings of males are vivid metallic green and yellow against a velvety black background and hindwings are greenish with bright yellow and black spots. As birdwings glide and flap their way through the forest their brilliance makes an outstanding splash of colour amongst the leafy greenery and shadows. These magnificent butterflies are commonly seen in rainforest when nectar-bearing flowers are abundant, particularly those near to **Aristolochia** vines, the food plant of their caterpillars. Most of the courtship dance is performed by the male birdwing. Hovering around and about his chosen mate while she continues to forage unconcerned by his attentions, he follows her from flower to flower occasionally gliding down in front of her to display his colour. Once the nuptial dance is complete they both fly off to nearby leaves where they land and mate. The female then lays her eggs on **Aristolochia** upon which vine the caterpillars will feed and often completely defoliate.

The only other North Queensland rainforest butterfly to match the size and splendour of the Cairns Birdwing is another protected papilionid, the magnificent blue Ulysses. The startling brilliance of the vivid blue wing markings of this species provide an impressive sight. When sunlit they give the impression of a series of successive bright blue flashes moving through the dark vegetation.

Photo   :   Paluma

COMMON EGGFLY BUTTERFLY
Hypolimnas bolina

Another colourful butterfly commonly seen in tropical Queensland, often along rainforest edges, is the Common Eggfly. Although many butterfly species forage among canopy blossoms some, that are not necessarily forest dwellers, visit the flowers of invading weed plants such as those of lantana (right, top). Lantana was originally introduced into Australia from Central and South America as a decorative garden plant, but rapidly became a major pest. In areas of forest that have been cleared by forestry activities or devastated by cyclones, Lantana is one of the first plants to colonise the open sunnier areas. Whilst it is an extremely troublesome invader, its bright pink, yellow, orange or red flowers attract many butterflies, such as this Common Eggfly, and its berries are a favourite food of crimson rosellas and other fruit-eating birds.

Photo   :   Paluma

BIRD-EATING SPIDER
Solenocosmia crassipes

Bird-eating spiders build loose sheet-like webs under fallen logs or, like the one illustrated here, under bits of tin sheeting or other debris close to homesteads. Closely related to the infamous Australian funnel web spiders, these huge furry brown bird-eaters, almost the size of a mouse, are known to eat animals such as mice, birds, and frogs. Like all spiders they are venomous and inject a poison into their prey to quickly immobilise it before feasting upon it.

As well as many different kinds of spiders, which of course are not insects but are technically termed arachnids, the forest floor abounds with a vast array of small invertebrate creatures such as leeches, scorpions, millipedes, centipedes, slugs and snails. Hidden away amongst leaf litter, mosses, rotting wood, soil and beneath bark, these animals are important foods for forest floor dwelling frogs, skinks, lizards, birds and mammals.

Photo : Iron Range

AMPHIBIANS

The amphibians are best known to most people as frogs and toads. Two lesser known groups of amphibians are:- (a) the newts and salamanders, about 300 species found predominantly in the northern hemisphere; and (b) about 80 species of strange worm-like 'legless frogs' called caecilians which are confined to the tropics excluding Australasia.

Australia has some 170 frogs, only a small proportion of about 2500 on earth; a situation doubtless due in part to the general aridity of Australia.

CANE TOAD
Bufo marinus

No true toads occurred in Australia until 1935 when the, now infamous, Cane Toad (below) was introduced at Gordonvale, North Queensland from Hawaii in the hope it would destroy beetles damaging sugar cane. This highly successful amphibian rapidly expanded its geographical range at 8.1% annually, to now be present over 33.8% of Queensland, or 584000 square kilometres. It is now on the New South Wales border and is getting closer to the Northern Territory. This introduced giant amphibian has caused great trouble to native Australian fauna, and to man. The toad is large enough to swallow whole native animals such as insects, amphibians, reptiles and mammals. It secrets a poison from the large gland behind each eye which can kill dogs and other animals. When these toads

jump into domestic fowls' drinking water they can cause serious economic problems. We include this creature as one example of what can happen when man interferes with the natural balance of things by introducing an exotic animal. The picture shows a mating, or amplexing, pair.

Photo : Paluma

NORTHERN GREAT BARRED FROG
Mixophyes schevilli

This spectacular frog (opposite,top), with a maximum body length of 90 millimetres, inhabits rainforested mountain streams of the Atherton Tableland and south to Paluma. Its large powerful hind legs enable it to leap considerable distances over the forest floor when disturbed. Huge eyes with a dark iris give 3 of the 4 great barred frogs a most appealing appearance.

Photo : Paluma

TREE FROG
Nyctimystes tympanocryptis

The 3 or 4 lovely Australian **Nyctimystes** tree frogs are well confined to areas about the Atherton Tableland, where they live in rainforest. They too have large dark eyes, making them most attractive. The species illustrated (right, bottom) is the most common of this little known group of frogs.

Photo : Mount Bellenden Ker, coastal slopes

WOOD FROG
Rana daemeli

All other continents boast a good number of the so-called 'true' frogs of the family Ranidae, but Australia has only one which is the Wood Frog (above). Found in Northern Queensland from the tip of Cape York south to the Townsville area on the east coast, and to the Mitchell River and perhaps as far south as Normanton on the west coast it inhabits rainforest, paper-bark swamps and other habitats close to streams and lagoons where it is active, vocal, and predacious at night. Note the blood-gorged mosquito feeding above the frog's eye.

Photo   :   Iron Range

GREEN-EYED TREE FROG
Litoria serrata

A fine example of adaptive colouration is the Green-eyed Frog, a tree frog that has developed colouration blending with its perching situations of moss and lichen covered branches close to rainforest streams. Moreover, it has peculiar 'frills' of skin along its limb edges to break up the profile and make it more difficult to see. Its six or so short knocking call notes can be heard at night in rainforests from Paluma to the Atherton Tableland, and north to an area north of Cooktown.

Photo   :   Paluma

RED-EYED TREE FROG
Litoria chloris

One of our most spectacular tree frogs is the Red-eyed Tree Frog, which is very rarely seen except during and immediately after heavy spring and summer rains, when individuals quickly congregate in low shrubs and grasses. In these brief nocturnal gatherings they continuously emit long 'moans' usually followed by a lovely soft trilling sound. These vocalizations are accompanied by an impressive inflation of the yellow vocal sac which is presumably also a visual display. Toe pads on the tip of each digit, so typical of tree frogs, are effectively 'sucker pads' much assisting these frogs to climb. This frog is found only on the east Australian coast from the lower part of Cape York to Gosford district near Sydney, in rainforest and other wet forests, floodplains and coastal rivers.

Photo   :   Paluma

GIANT TREE FROG
Litoria infrafrenata

Perhaps Australia's largest frog, growing to a body length of 140 millimetres, is the Giant Tree Frog. Any meeting with this spectacular amphibian is memorable but we particularly remember this photograph. Camped in Iron Range forest we reached a creek edge one night to draw water. In our torch beam at eye level we suddenly saw this fine individual, at close to maximum size for its species. Dawn held the beam still while Cliff rushed to camp for cameras and flashes. How relieved we were to receive the processed films recording this chance encounter. The Giant Tree Frog occurs from about Cardwell northward to the tip of Cape York on the east coast; and down the west Queensland coast to Weipa, and possibly Normanton. It is also throughout New Guinea and surrounding islands and is not restricted to rainforest but might be found in many other habitats.

Photo   :   Iron Range

REPTILES

Australia is rich in reptile life, with some 2 crocodiles (both of which can be encountered in lowland rainforest creeks), 6 sea turtles, approximately 16 freshwater turtles, 65 geckoes, 90 lizards, 21 monitors or goannas, 100 skinks (simply more lizards to the layman's eye), 110 land snakes and 32 sea snakes. Australia is well known as home to many dangerous snakes such as the Taipan, Brown and Tiger snakes. Few dangerous snakes occur in our tropical rainforests, however. One exception is the very rarely encountered Rough-scaled Snake **Tropidechis carinatus** which is very aggressive when disturbed and is potentially very dangerous. In our five years intensive rainforest work at Paluma we have only seen one Rough-scale. The Red-bellied Black Snake (page 37) is a common venomous species in tropical rainforest, but is not regarded as a deadly snake.

SAW-SHELLED SNAPPING TURTLE
Elseya latisternum

This is one of those fresh water turtles, or tortoises as some would prefer, which can be seen in Australian tropical rainforest creeks, rivers and lagoons. Members of the snapping turtles have a characteristic horny shield or cap over the top of the skull which develops early in life. Their jaws are very strong and sharply edged and these are used to snap at prey, such as fish and frogs, hence the

name. They can be aggressive when handled and are capable of inflicting a nasty bite, or deeply tearing hands with sharp claws which they use to tear food apart. This individual (above) paid us a visit one night by swimming into our bathing area!

Photo : Iron Range

NORTHERN LEAF-TAILED GECKO
Phyllurus cornutus

Considered Australia's largest gecko the Northern Leaf-tailed may attain a total length of 25 centimetres. It is a classic example of adaptive body shape and colouration, resulting in a most cryptic creature (opposite). Pigmented in natural browns and greens and with a flattened and irregular outline this gecko, when pressed against its tree trunk microhabitat, is almost invisible because its flatness eliminates even the smallest of shadows. Even the iris is camouflaged, by irregular pigmentation, to add to the overall difficult-to-see effect (page 30, top). It is a night-time forager, seeking out its predominantly insect foods on tree trunks and boughs and on the leaf litter of the forest floor. During daylight hours it shelters beneath tree bark for which it is of course ideally adapted by its flat profile. Unlike the climbing geckoes adapted for smooth surfaces with pads on their digits this gecko is clawed for better traction on rough surfaces. It occurs in rainforest and adjacent forests, and wet sclerophyll forests from eastern Cape York Peninsula to northern New South Wales.

Photo : Paluma

GIANT TREE GECKO
Pseudothecadactylus australis

Officially this magnificent and rare gecko (below) has no common name but as its, slightly smaller, closest relative is the Giant Cave Gecko we have given it the above name for convenience until a common name is formally established. It has only been recorded on large tree trunks so the name is applicable to this agile climbing gecko which is truly remarkable in having modified scales beneath the tail tip to form adhesive lamellae just as it and many other climbing geckoes have under their toe pads. This reptile is confined to the north-eastern coastal ranges of far north Cape York. We spent eight full weeks in Iron Range rainforest eager to see this unique gecko and we were thrilled when, on our second to last night in the area, one literally fell or leapt from a tree above our camp onto our Landrover bonnet!

Photo   :   Iron Range

BOYD'S RAINFOREST DRAGON
Gonocephalus boydii

One of three closely related rainforest dwelling dragon lizards is the handsome and rarely seen Boyd's Rainforest Dragon (right); which is confined to the forests of tropical North Queensland from Paluma northward. It is a slow moving lizard, mostly of the tree trunks and forest floor which relies on immobility and cryptic colouration to avoid detection. Most people in fact see this lizard on rainforest roads. The loose colourful skin hanging below the jaw is know as a gular pouch, or dewlap, which many lizards of the family Agamidae have, and use by raising them in display.

Photo   :   Paluma

EASTERN WATER DRAGON
Physignathus lesueurii

Another member of the lizard family Agamidae, the Eastern Water Dragon (right), is far more widespread than the Boyd's (page 31). It is found in various habitats close to water, including the coastline, and extends from north of Cooktown down the east coast of Australia into eastern Victoria. This large semi-aquatic lizard, attaining almost a metre in length, is commonly seen above or beside water perched on tree or rock from which it will drop if approached. Hitting the water the lizard submerges and swims away to safety.

Photo : Paluma

SPOTTED TREE MONITOR
Varanus timorensis

Australia's 21 monitors, also known as goannas, represent over two thirds of the world's 30 species of these huge lizards. The largest living on earth is the Komodo Dragon of Indonesia which grows to more than 3 metres in length. Australia's largest is the Perentie reaching a mere 2.5 metres in length! Monitors have very long forked tongues which flicker in and out to 'smell' the environment. The Spotted Tree Monitor is one of our smaller species, reaching only 0.6 of a metre in total length. It is found throughout the north of Western Australia, Northern Territory and Queensland, north of the Tropic of Capricorn. It is a tree living species, not restricted to rainforest, which shelters in tree hollows, holes, or beneath tree bark. It eats insects, lizards and nestling birds.

Photo : Paluma

BURROWING SKINK
Anomalopus frontalis

Unlike the more typical skinks (below) some species have, in evolutionary terms, entirely lost their legs as an adaptation to a burrowing life-style in which limbs would not only be unnecessary but a distinct hindrance. These legless skinks feed on insects and other invertebrates in the leaf litter, soil, or beneath bark etc. As a result of their lack of legs they are sometimes mistaken for snakes. Unlike snakes, however, they have a broad unforked fleshy tongue and can produce simple vocalizations. Of course they are completely harmless. **Anomalopus frontalis** is one of many burrowing skinks, but it is one of the least known being confined to tropical rainforests of the Atherton Tableland area southward to Paluma where it is found under fallen timber or in decaying logs.

Photo   :   Paluma

EASTERN WATER SKINK
Sphenomorphus quoyii

This lovely skink has an extensive distribution from North Queensland in the Cairns area southward down eastern Australia to the Victorian border and then south-west in a narrow belt to Adelaide in South Australia. It commonly associates with water and will not hesitate to enter it to swim to an opposite bank to avoid disturbance. The skink photographed is resting on a fallen bough covered with unusual tiny orchids of the **Bulbophyllum** group.

Photo   :   Paluma

GREEN PYTHON
Chondropython viridis

The pythons include many of the largest and most long-lived snakes. The Reticulated Python of India, South East Asia, Indonesia and the Philippines may attain over 10 metres in length. This huge serpent, the Indian Python and the Anaconda of South America, may exceed 200 kilograms in weight and could kill a man by constriction which is how all pythons kill prey; none of them being venomous. Captives have lived more than 25 years, and may well be able to survive half a century in the wild. Most pythons lay eggs which females care for by remaining coiled around them until hatched — primitive birds really!

Of Australia's dozen pythons the Green is undoubtedly the most rare and exciting, being a rainforest species confined to the extreme far northern areas of Cape York where it is rarely encountered, and reaches nearly 2 metres in length. We allowed eight weeks to find this elusive photogenic subject in Iron Range forests, but were incredibly lucky in finding one on our third night! We did not see another. This individual has just gained green colouration; younger snakes being a bright-yellow or rust-red. Pits along the lower jaw and on the snout contain heat-sensitive tissue with which the snake detects warm-blooded prey.

Photo : Iron Range

CARPET PYTHON
Morelia spilotes

Carpet pythons may reach 4 metres in length and are so named because of their lovely reticulated colourful patterns. This python is distinguished from the otherwise superficially similar Amethystine Python (below) by the fine granular scales on top of the head as opposed to the large plate-like scales on the Amethystine. This snake has a vast distribution, being found throughout the continent, except the south east and Tasmania, where it occurs in many habitats. There are two basic colour forms, the more distinctly blackish southerly one being called the Diamond Python. The individual shown here is a lovely 'golden' form from the North Queensland hill forest and was photographed as it lay curled on the leaf litter.

Photo : Paluma

AMETHYSTINE PYTHON
Morelia amethistina

Australia's largest snake, the Amethystine or Scrub Python has been recorded at 8.5 metres, or almost 28 feet long; being about 3.5 metres long on average.

It is a slender python however and it is not, therefore, a huge serpent like some of those of Asia, Africa and South America. This rainforest snake is found along the North Queensland east coastal areas from Paluma north to the Cape York tip; and also on off-shore islands, those of the Torres Strait, and on New Guinea. It lives in rainforest and more open habitats and is called the Amethystine Python because of the lovely hues of colour caused by refraction of light in its scale structures. The snake pictured was a little over 4 metres long, and is just starting to uncoil from is resting posture on the forest floor. Its food includes pademelons (small wallabies), fruit bats and other mammals.

Photo   :   Paluma

BROWN TREE SNAKE
Boiga irregularis

This snake (below) is not to be confused with the more slender, smaller-headed and harmless tree snakes (see below) as it is venomous, although not considered dangerous to humans.

Usually brown, or pale with russet or brown barring, it is readily identified by the bulbous head and huge eyes. Up to 2 metres long, it is nocturnal and tree-dwelling, eats birds, lizards and small mammals, and shelters in tree or rock crevices, caves or beneath stones in daytime. It is found throughout northern and eastern Australia as far south as about Sydney in many habitats including tropical rainforests.

Photo   :   Paluma

NORTHERN TREE SNAKE
Dendrelaphis calligaster

This is one of two slender, agile, tree-frequenting snakes with long whip-like tails growing to 1 to 2 metres long. These attractive and colourful snakes exhibit considerable colour variation through their ranges. The Northern Tree Snake is confined to the east of tropical North Queensland from about Townsville north to the tip of Cape York, Torres Strait Islands and New Guinea. It is solid toothed and non-venomous, hunting various foods including birds, lizards and frogs mostly in trees and shrubs but also on the ground, where it often suns itself.

Photo   :   Iron Range

RED-BELLIED BLACK SNAKE
Pseudechis porphyriacus

This front-fanged venomous snake is very closely related to some very dangerous Australian snakes. It is not, however, considered very dangerous to adult people but a good bite from a large snake might kill a child. Like almost all snakes it is retiring by nature, very rarely striking a human unless provoked or handled. It is found from southern Cape York Peninsula southward down the east coastal area to Melbourne and west almost to Adelaide, in various habitats. Very large individuals reach over 2 metres in length. Food consists of frogs, small mammals, fish, lizards and snakes.

Photo : Paluma

BIRDS

Of approximately 9000 species of birds on earth some 720 are found in Australia; of these roughly 130 typically occur in tropical rainforest and about 35 of them are entirely confined to this habitat.

AUSTRALIAN CASSOWARY
Casuarius casuarius

The Australian Cassowary is one of the Ratites, the most primitive group of living birds on earth, consisting of two other cassowary species in New Guinea, the Emu, the Ostrich, two Rheas and three Kiwis. All are large to huge flightless birds. Cassowaries are tropical rainforest fruit-eating birds and the Australian one is found in the forests from Paluma to the top of Cape York, and in New Guinea. All ratites have rather coarse dense plumage and almost completely lack wings, having only the vestigial remains of these limbs. Their legs are huge and powerful relative to the rest of their body. This is particularly true of cassowaries which if cornered and harassed will kick out in defence, and may kill humans this way. The huge toes are tipped with dagger-like nails that can tear flesh apart and several people in North Queensland, and many more in New Guinea, have been badly wounded or even killed by them. Thus, one of the potentially most dangerous animals in Australian tropical rainforests is a bird!

As in some other ratites, the male cassowary appears to incubate the eggs and care for the young once the female lays her eggs in his nest, and males are most aggressive when caring for their young. The huge eggs are a lovely blue-green and number 4 or 5. Chicks (above) are delightfully striped, rather like young wild pigs, and this obviously makes them more cryptic on the forest floor. Other than their dense blue-black feathering adult cassowaries sport bright hues of red and blue bare skin about the head and neck with bare red neck wattles. Young birds (right) start to develop a horny crown atop the head called the casque which becomes large and erect in adult birds (opposite). This casque apparently protects the huge bird's head as it rushes through the dense vegetation head first. These birds may weigh up to 55 kilograms and stand as high as an average woman.

Photos : Paluma, and Papua New Guinea (adult)

BRUSH TURKEY
Alectura lathami

This bird (above) is one of the three Australian megapodes, or mound builders. Megapode means large foot, and these birds have large powerful legs and feet for scratching leaf litter and soil in search for food. Males also use them to build the huge nest mounds that these birds are well known for. Megapodes do not incubate their eggs like other birds, but the females lay them deep in the male's mound of accumulated vegetation scratched from the surrounding area. This 'compost heap' generates considerable heat as it ferments and this incubates the eggs. The male critically regulates both temperature and humidity by adding or removing material, and possibly measures these conditions with sensitive areas of facial skin or mouthparts as he thrusts his head into 'test holes'. When the eggs hatch the remarkable precocious young literally burrow upwards through the rotting vegetation and burst out into the wide world to fly and scurry off to fend for themselves.

Photo : Paluma

RED-NECKED RAIL
Rallina tricolor

A truly rainforest bird the Red-necked Rail (opposite, below) is restricted to tropical forests from the northern extremity of Cape York southward to Paluma. It is also found on New Guinea and surrounding islands, and in the Moluccas. Like so many of the rails it is a very difficult bird to see normally, and all the more so because of its dense habitat. It can often be heard, however, giving its rather monotonous repetition of loud short notes at dusk and dawn. Younger birds lack the lovely orange-brown head, neck, throat and breast and are a uniform nondescript colour the same as the body of the adult. This retiring and secretive bird is most frequently encountered along water courses where it moves silently about mud and shallows or creek-side vegetation seeking out its predominantly freshwater animal foods.

Photo   :   Paluma

PURPLE-CROWNED PIGEON
Ptilinopus superbus

Of 22 Australian native pigeons 7 might loosely be termed tropical fruit pigeons. Certainly the most colourful, and arguably the most splendid of these, if not of all Australian pigeons, is the Purple-crowned.

This bird is found throughout our tropical rainforests from the Cape York tip southward down the east coast to near Rockhampton, Queensland, but it is uncommon south of Paluma and is more numerous in the lowland forests than on the ranges. The black pigmentation on the iris directly before the pupil helps break the eye up to make it less conspicuous. Notwithstanding this bird's brilliant colouration it is extremely cryptic in its leafy environment and is most difficult to spot unless moving or calling. It builds a particularly flimsy nest, although all pigeons' nests are slight, being no more than a few fine sticks precariously placed on a branch fork, upon which the single egg is laid. If one of the pair is flushed off the nest the egg or young can usually be clearly seen through the sparse nest from directly below. Most tropical fruit pigeons are very important dispersal agents for rainforest trees and plants. They swallow whole soft succulent fruits, digesting the rich pericarp and passing undamaged seeds. In this way the seeds are often voided well away from the parent plant and in a suitable situation, thus much improving the seeds' chances. Moreover, experiments have shown that some plant seeds are more viable having passed through such birds.

Photo   :   Paluma

DOUBLE-EYED FIG PARROT
Cyclopsitta diopthalma

Australia's smallest parrot, with three distinct isolated populations; the picture being a male of the most northern form known as Marshall's Fig Parrot, which is confined to the Iron Range forests. Females in this area lack red about the head, which is replaced by blue hues. Their name refers to their basic diet of fig fruits, and their feeding habit of tearing into such fruits is clear in the photograph. They breed in isolated pairs, excavating a small hole and cavity in rotten wood. Two white eggs are laid in the nest hole which the female incubates, and when they hatch both sexes feed the tiny young.

Photo   :   Iron Range

CRIMSON ROSELLA
Platycercus elegans

The well known popular Crimson Rosella is widespread down the Australian east coastal areas from southern Queensland to south-eastern South Australia and Kangaroo Island; and an isolated population, of smaller darker birds, occurs from Atherton Tableland south to the Eungella Range, Herberton and Ravenshoe area. This northern population is mostly in tropical and subtropical rainforest, but southerly birds occur elsewhere. Mostly a seed eater, it also takes fruits, berries, buds, shoots, blossoms and insects.

Photo   :   Fraser National Park, Victoria

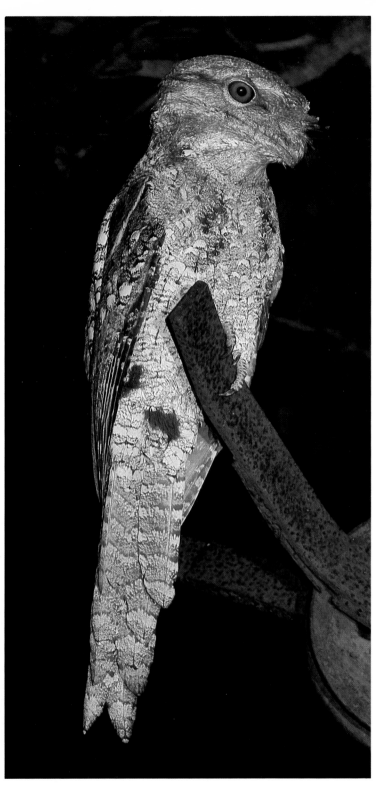

PAPUAN FROGMOUTH
Podargus papuensis

Frogmouths are commonly mistaken for owls which they superficially resemble and are closely related to, but they have weak small feet, not talons, and a huge mouth which earns them their common name. The great bill is used to snap up animal prey from the ground or trees at night. Like owls, and nightjars, their plumage is soft and highly modified to produce very little or no sound in flight. Thus, their unsuspecting prey fails to hear their swooping attack. The huge Papuan Frogmouth at 45 to 58 centimetres in length is the largest of the three Australian species and has a discontinuous distribution from Townsville north to the tip of Cape York where it is found predominantly in tropical rainforest and other dense growth. Like other frogmouths it sits motionless in trees during daylight in a posture that makes it appear as a piece of tree bough. As its name implies it also occurs in New Guinea and surrounding islands. Our photograph was taken of a hunting bird perched on the apex of an old crane dumped at the edge of the forest.

Photo   :   Paluma

TAWNY FROGMOUTH
Podargus strigoides

This frogmouth (page 44) is far better known than the last as it occurs throughout most of Australia, even in the arid centre, and can be seen in all major city suburbs. It differs from the Papuan in being smaller, having a relatively shorter tail, and a predominantly yellow, not red, iris.

Photo   :   Paluma

LITTLE KINGFISHER
Ceyx pusillus

Smallest of 10 Australian kingfishers is the Little (page 45), which is one of only two of our birds that typically feeds by diving into the water for prey. Most of our kingfishers feed over land. The Little Kingfisher is confined to tropical North Queensland and the Northern Territory tropical coast, being apparently absent from the southern Gulf of Carpentaria, and is also in New Guinea, the Solomons and the Moluccas. It inhabits mangroves and rainforests where it digs burrows into stream banks to nest.

Photo   :   Paluma

AZURE KINGFISHER
Ceyx azureus

Second smallest Australian kingfisher and the other 'true' kingfisher that feeds by diving into water for food is the colourful Azure (opposite, top). Its extensive distribution is from the western Kimberleys of Western Australia across the northern tropics to Cape York and right down the east coast to Victoria, Tasmania and Adelaide in South Australia. It frequents forested streams, including rainforests, and mangrove waterways, lakes and swamps.

Photo : Paluma

WHITE-TAILED KINGFISHER
Tanysiptera sylvia

Undoubtedly our most spectacular kingfisher, having two central white tail feathers almost as long as the rest of the bird which trail delicately behind it when flying. The bird (opposite, bottom) migrates from New Guinea in November to breed in tropical rainforest north from Paluma, predominantly in lowlands. It nests by burrowing into termite mounds on trees or the forest floor and feeds on insects, frogs and lizards. By late April or early May these birds have returned to New Guinea.

Photo : Paluma

NOISY PITTA
Pitta versicolor

Four of the world's 28 pittas occur in Australia. They are the most highly coloured of our rainforest floor birds. The Noisy Pitta (below) is our largest species and inhabits the tropical and sub-tropical rainforests from the extreme northern tip of Queensland almost to the Sydney area. It is one of quite a few rainforest birds that specialize in living on the forest floor leaf litter animal life; as do the four birds on the following two pages. Pittas conspicuously hop about the floor and on fallen logs in search of their predominantly earthworm and snail diet, but also eat insects and some fruits. This lovely bird is eagerly sought out by bird watchers as one of our most spectacular species, but it is more easily heard than seen. Its clear powerful whistles clearly resemble someone whistling 'walk-to-work' repeatedly, and birds can be attracted to the observer imitating this call. Part of our Noisy Pitta population migrates to and from New Guinea whilst others are clearly permanent residents, but this aspect of the bird's biology requires further examination. The large bulky nest is built on or very close to the ground and is a domed chamber of large sticks and other material, with a side entrance hole. Often a 'platform' or 'door-mat' of sticks is laid in front of the entrance hole which for some unexplained reason birds frequently decorate with animal dung. The picture shows the parent bird on its way to the nest to feed its young some worms.

Photo : Paluma

BASSIAN THRUSH
Zoothera lunulata

A widespread population of birds previously known as White's or Scaly Thrush, distributed from the Atherton Tableland down the Australian east coast to Victoria, Tasmania and eastern South Australia, was in 1983 demonstrated to include two different species. The Bassian Thrush occurs in two forms, one in South Australia, Victoria and Tasmania, eastern New South Wales and south-eastern Queensland and another, the one illustrated, in the highland rainforests of the Atherton Region. The other species, now known as the Russet-tailed Thrush, *Z. heinei*, occurs in north-eastern New South Wales and south-eastern central coast and north-eastern Queensland where it meets the Bassian Thrush. The two species have differing calls and the Russet-tailed Thrush has a shorter tail, rufous-tinged rump, and russet tail but they are nevertheless very similar in appearance. The Bassian Thrush is a shy ground-frequenting thrush that feeds singly or in pairs on insects, worms, other invertebrates and fruit.

Photo   :   Paluma

NORTHERN LOGRUNNER
or
CHOWCHILLA
Orthonyx spaldingii

One of only two logrunners, the Chowchilla is restricted to tropical rainforests from about Cooktown to Paluma and is more abundant in the highlands. It was also known as Spalding's Spinetail, for this stocky leaf litter feeding bird has a tail modified into strong spiny shafts upon which it props itself whilst leaning backwards and vigorously throwing litter aside with one foot in its search for animal foods. It moves rapidly about the dim forest floor in flocks of about 3 to 6 birds within the flock territory, and possibly the flock members are closely related to each other. They are very vocal birds and their loud dawn and dusk chorus is characteristic of the northern upland rainforests and the quality of their strong song is reminiscent of lyrebirds calling. Males have a pure white breast and females a rich brown-orange one, sometimes with a small patch of white below. The photograph depicts a female just about to leave the bulky stick and moss domed nest with a nestling faecal sac.

Photo   :   Paluma

PALE YELLOW ROBIN
Tregellasia capito

This appealing little robin is a bird of both tropical and sub-tropical rainforests and it is most commonly found where lawyer vine is common. It ranges from just south of Cooktown to the Paluma area and from about Rockhampton to near Sydney. It forages over forest foliage in the understorey during the warmer months but in dry winter months it more often feeds closer to or on the ground, when it behaves rather like the Grey-headed Robin in resting on vertical perches from which it pounces to the ground after a worm or insect. In very cold and dry months they become common garden birds along the forest edge. This bird often nests on armed lawyer vine stems, as do others (see below).

Photo   :   Paluma

GREY-HEADED ROBIN
Poecilodryas albispecularis

This lovely robin is typically a bird of the hill rainforests from near Cooktown southward to Paluma and is usually seen singly or in pairs. Its lovely tortoiseshell colouration is striking close-up, but very effectively breaks up the bird making it difficult to locate when it is still and at a distance. It feeds actively on the leaf litter of the forest floor, typically surveying the ground from a vertical sapling trunk and fluttering down to snatch or forage after animal prey. Nesting Grey-headed Robin pairs build their delicate tendril and moss nests on branch forks from 1 to 4 metres above the ground, very often on the spike stems of lawyer vine (see page 8). The male often provisions the brooding female by flying close to the nest and calling her off to offer her his insect, worm or small lizard foods. At such times a lovely raised-wing display is given as is shown in the photograph.

Photo   :   Paluma

GOLDEN WHISTLER
Pachycephala pectoralis

By no means confined to tropical rainforest, the Golden Whistler occurs in the denser vegetations of temperate rainforests, woodlands, riverine forests and denser inland habitats. It ranges from a point near Cooktown in North Queensland right down eastern Australia through Victoria and an extensive area about the Adelaide region of South Australia to south-eastern and southern Western Australia. It is a well known and loved bird because of the male's brilliant sulphur-yellow and olive-green body plumage with white throat and black head markings. The female is drably garbed in olive-greys and very pale yellows. The strong, clear whistling call notes of this bird add to its appeal.

Photo   :   Paluma

BOWER'S SHRIKE-THRUSH
Colluricincla boweri

Like a number of North Queensland rainforest endemic birds this species has a very restricted range on forested mountain ranges from just south of Cooktown to Paluma higher than about 400 metres above sea level. The shrike-thrushes are very closely related to the whistlers and are in fact little more than drab whistlers, and are also fine singers. Bower's Shrike-thrush typically produces a loud 'chuck' note followed by clear whistling notes and is usually heard and seen as single birds or in pairs. It is not an easy bird to see as it is loathe to permit a close approach, and is often confused with the Rufous Shrike-thrush which is in fact easily distinguished by its flesh coloured bill as opposed to the black bill of Bower's Shrike-thrush. It is an active foliage-gleaner taking insects from leaves and vine tangles, and commonly tearing apart dry dead vegetation.

Photo   :   Paluma

MACLEAY'S HONEYEATER
Xanthotis macleayana

Yet another bird of restricted range from Cooktown south to Paluma; on the rainforested ranges and rainforest, woodland and riverine vegetation of the lowlands. This is one of Australia's 68 honeyeaters, a family of 170 species in the Australasian region, a family which forms one of Australia's characteristic bird groups. As the name implies, they frequently feed on flower nectar; and many species are as important to native flowering plants as pollinators as are insects. In rainforest Macleay's Honeyeater is not an easy bird to see well because it is quieter than many honeyeaters. A particularly good place to see this bird, however, is the tea-house garden in Paluma village where the species comes in large numbers to feeders, and will even feed from the hand.

Photo : Paluma

BRIDLED HONEYEATER
Lichenostomus frenatus

This is primarily a bird of the tropical highland rainforests, although it can be seen at some lowland rainforest locations such as Cape Tribulation. It also occurs in wet forests along water courses, swamp woodlands and in drier forests adjacent to rainforests. Like many honeyeaters it is very active, vocal and aggressive. Birds can sometimes be seen pressing home vigorous attacks upon another, and birds gripping each other and tumbling down to bump onto the forest floor are not infrequently observed in summer. This bird is particularly fond of taking nectar from flowering climbing pandanus (see page 9) and the Umbrella Tree **Schefflera**. This honeyeater's complex facial and bill colouration can be clearly seen in this head portrait.

Photo : Paluma

BOWERBIRDS

Bowerbirds are world renowned for the remarkable behavioural developments of males of most species; which reaches a peak in those that build and decorate bowers — stick structures that are constructed and maintained by dominant males as an asset in attracting potential female mates and perhaps deterring rival males. Some males such as the Satin Bowerbird even 'paint' their bower walls by applying pigmented fruit pulp and other matter to bower sticks with the aid of a fibrous wad or 'brush' held in the bill tip. Few, if any, other animals construct what is effectively a symbolic structure. The bower is purely a male-made phenomenon having nothing whatever to do with the nest, or nesting, which is a typical cup-shaped nest built in a tree or crevice by the female away from the bower. Bower-building males play no part in nest building, egg incubation, or raising of young; which are performed solely by females once mated by the male of their choice at his bower. It has been noted that males of species with the most elaborate bowers have generally dull plumage whereas those with simple bowers have bright plumage and ornate crests or capes of long feathering. Thus a 'transferral' effect can be inferred, with males losing their bright sexual characters of colourful plumage as evolving more complex bower structures instead.

So remarkably complex, large, and highly decorated are some bowers that early explorers to the homelands of these birds in Australia and New Guinea at first refused to accept them as bird-built, but were convinced they were constructed by native peoples for their children's entertainment. Of the 18 kinds of bowerbirds 9 are found only in New Guinea, 7 are confined to Australia and 2 occur on both land masses. Most are tropical rainforest birds from lowlands up to 4000 metres above sea level, although several have adapted to arid and semi-arid Australian conditions.

Males spend much time maintaining their bowers and calling about them, in order to attract females and deter rivals. They will fertilize many females in a season and are thus promiscuous. Females, having been mated by the male of their choice are kept busy for the rest of the season sitting on their 1 to 3 eggs and raising the resulting young. It is generally thought that the abundant food resources of these promiscuous rainforest bowerbirds enables them to have sufficient time and energy required for males to spend much time in bower maintenance and attendance, and the females to raise the young unaided. So well known is the male bowerbird habit of bringing decorations to bowers that Australians refer to a person who is loathe to part with anything as a 'bit of a bowerbird'.

GOLDEN BOWERBIRD
Prionodura newtoniana

The world's smallest bowerbird, about the size of a Common Starling, is the Golden Bowerbird which builds the largest of bowers. Its typically twin towered, bower structure can be 3 metres tall and constitutes a considerable mass of sticks. The male (opposite, top) is brilliant golden-yellow and its unique feather structure refracts light often producing pure white highlights on the plumage. Females however, are dressed in greys and olive-browns doubtless to render them far less conspicuous whilst raising the all important 1 or 2 young in the small cup nest in a tree or crevice (below).

Adult males probably do not attain their golden plumage until the 6th or 7th year, as in the Satin Bowerbird, and until then have the appearance of females. Their bower usually consists of 1 or 2 towers of sticks with a display perch protruding from a single tower bower, or connecting between the 2 towers (opposite, bottom). The sticks of the towers meeting this display perch are more skillfully and regularly laid and it is on these that decorations of white, off-white and pale green orchids, other flowers, seed-pods and lichens are placed.

Golden Bowerbirds have a very restricted range, being confined to tropical rainforests above about 900 metres altitude from just south of Cooktown to Paluma. Many people travel very considerable distances both interstate and internationally, with a specific desire to see this bird and its magnificent bower, particularly in the Atherton Tableland and Paluma areas. The best time to see them is, unfortunately, the wetter months of November to January when males are active at bowers and females are nesting. Whilst many interested persons see males at bowers very few indeed have ever seen a female at the nest; for the bird's environment is most unpleasant at nesting time due to heat, rain, mosquitoes and leeches. It is, however, the period of greatest abundance of the insect and fruit food required for raising the young.

Photo   :   Paluma

SATIN BOWERBIRD
Ptilonorhynchus violaceus

Best known of all the bowerbirds, because of its extensive distribution down eastern and south-eastern Australia from Cooroy, south-east Queensland to the Otway Ranges, Victoria, including the Sydney and Melbourne areas, is the Satin. A population of smaller birds of this species occurs in tropical North Queensland from the Atherton Tableland district south to Paluma. Unlike the 'maypole' type bower of the Golden, the male Satin builds an 'avenue' bower of two parallel walls of sticks decorated with blue or green objects (above). This preference for 1 or 2 colours is typical of many bowerbirds. As in other promiscuous bowerbirds males attract many females to the bower for mating and the females (right) raise the young alone. When not breeding Satin Bowerbirds, unlike the Golden, form flocks which can number several hundred birds. At such times they frequently leave rainforests and denser woodlands and feed in open areas, even feeding on grasses in large numbers. Food is predominantly fruits, leaves, animals, seeds and flowers; but females raise their young almost entirely on insects.

Photo : Paluma

TOOTH-BILLED BOWERBIRD
Scenopoeetes dentirostris

The mysterious Tooth-billed Bowerbird (above) is found over the same rainforested area as the Golden Bowerbird, but is slightly less restricted altitudinally, being found over 500 to 600 metres above sea level. Males are doubtless promiscuous but they do not build a structure, but clear an area of forest floor of all debris to form a 'court' upon which they place leaves upside-down to expose contrasting pale undersides (left). Here they call continuously through the display season of September to January, producing bewildering vocalizations including excellent mimicry of numerous other birds and rainforest sounds. In summer the abundance of calling birds leads one to consider this fascinating bird the most abundant of species, but in winter it is difficult to see or hear a single individual from one week to the next. At this time they are inactive and quiet in the canopy, living on fruits, leaves, stems and buds. The nest of this species is one of the most difficult of Australian birds to locate, being a very frail cup structure placed high in suspended vine tangles. As a result nothing is known of its nesting biology other than it lays 2 pale eggs.

Photo : Paluma

SPOTTED CATBIRD
Ailuroedus melanotis

The Spotted Catbird is one of two Australian bowerbirds that do not build bowers. Nor are they promiscuous breeders but, like the vast majority of birds, they form pairs and both birds assist in raising their young. Moreover, Spotted Catbirds appear to maintain their pair bond year after year; maintaining a breeding territory which expands and overlaps neighbouring pairs in the winter but which is vigorously defended and well defined when nesting. They eat predominantly fruits, especially figs, succulent leaves and shoots, insect foods and they even predate nestlings of other bird species. When raising their young they often take worms, from the forest floor, and are particularly eager to rob birds' nests of young which they tear apart to feed their own offspring. The nest is a bulky bowl-shaped structure of stout sticks and leaves placed from 2 to 10 metres above ground, usually in a sapling fork or infrequently in a lawyer vine (opposite). One to 3 eggs are laid during September to December.

The eggs hatch after about 24 days and the appealin young (below) fledge from the nest at about 22 days aft hatching. When approached a catbird pair with large youn in the nest, or with newly fledged young as illustrated, w perform impressive distraction displays. They will flutter the forest floor and, with wings held open and trailing on th ground as if broken, they stumble about in an attempt distract and divert one from their all important offspring. T one familiar with bird behaviour of this kind this trick doe not have the required effect but to the naive it is mo efficient as the observer, ignorant of young birds close b will follow the performing parent curiously. Potenti predators of the young birds are similarly distracted as th see what appears to be easy pickings in the form of a bad injured adult bird. If pressed closely by a predator, howeve a distracting' catbird quickly reverts to a healthy one an flees. Such behaviour is by no means unique to catbird but is very widespread in the bird world as a most successf subterfuge.

Photos : Paluma

VICTORIA'S RIFLEBIRD
Ptiloris victoriae

Named after the British Queen, Victoria's Riflebird is one of the fabulous birds of paradise for which New Guinea is particularly well known. Australia is, however, home of 4 of the 42 birds of paradise; 2 of them being other kinds of riflebirds and the fourth is the Trumpet Manucode of northern Cape York. The riflebirds like most bowerbirds and birds of paradise are promiscuous with the males (above) attracting many females (below) to their display and calling perches in the dense rainforest. Their penetrating loud calls advertise their presence to possible mates, and to rival males, and when a female arrives near their display branches they go into intense display to impress her. Throwing his peculiarly rounded wings up either side of the up-stretched head and neck the bird erects much of his lovely velvet and metallic plumage and sways slightly from side to side and bobs up and down while rapidly flicking his head first to one wing edge and then to the other. Occasionally the long bill is opened to reveal the brilliant yellow interior which is a startling and conspicuous feature in the darkness of the forest. Once mated the female has no more to do with the male but retreats to prepare for incubating her eggs and raising her young entirely alone. Australia totally lacks the woodpecker group of birds so familiar to people of Europe, Asia and America. The riflebirds, however, feed predominantly in rather woodpecker fashion. Having very powerful legs and claws they clamber over tree trunks and large boughs tearing tree bark and rotting wood in search of insects and insect larvae which constitute much of their diet. They also eat fruit. Like most bowerbirds, young male birds of paradise very closely resemble the females before attaining adult plumage.

Photos   :   Paluma

MAMMALS

All three groups of mammals; monotremes, marsupials and placentals, are represented in our tropical rainforests. Monotremes lay eggs, unlike the other two groups whose young are born live. Newly born marsupials are, however, embryonic-looking, and crawl into the mother's pouch where they develop further by suckling her milk. Young placental mammals do not require the protection of a pouch as they are retained within the mother for a considerable time gaining nourishment through her placenta. When born they cling to their mother to suckle, as do monkeys for example, or run free as do deer and antelope etc.

Many mammals of our tropical rainforests occur in no other Australian habitat and are found nowhere else on earth. Such mammals include some species of the appealing large-eyed ring-tail possums, the small primitive Musky Rat-kangaroo, and two species of tree kangaroo that, unlike their more familiar ground-dwelling counterparts, spend their lives in tree tops. Other animals such as the Striped Possum, Grey and Spotted cuscuses, the giant White-tailed Rat and the Cape York Rat, whilst being confined to the rainforests of north-eastern Queensland on the Australian continent, are also common in New Guinea.

ECHIDNA
Tachyglossus aculeatus

To the Europeans the Echidna looks like a large Hedgehog, or to Americans a small Porcupine; but there the resemblance, other than them all being mammals, ends. The echidna is covered in spiny quills as a means of protection. When bothered by potential predators, such as the Dingo, the Echidna rolls into a tight ball with its soft tube-like nose and clawed fleshy feet drawn tightly in to become almost impenetrable. The huge powerful claws of the Echidna enable it to burrow to safety, and to tear apart termite mounds and rotting wood to reach its predominantly termite diet which it eats with the aid of a long sticky tongue. This food and feeding method have given rise to its other common name of Spiny Anteater.

Few other animals are closely related to it except the well known and celebrated Platypus, which can also be seen in tropical rainforests swimming along creeks and rivers. the Echidna and Platypus are not confined to tropical rainforests. Echidnas, of which there is only one species in Australia, occur throughout Australia and New Guinea in a variety of habitats; and the Platypus, of which there is only the one species, is confined to aquatic environments along the eastern part of Australia. The Echidna and Platypus are the only two representatives in the world of the most primitive of the three mammalian groups, the monotremes (see above). The egg of the Platypus is incubated between the mother's body and tail which is folded forwards ventrally; and the newly hatched young cling to the mother's fur whilst suckling. Female Echidnas, like marsupial mammals, have a pouch. The Echidna egg is incubated in the pouch, so the young hatch directly into the mother's pouch where they then suckle milk.

Photo : Paluma

HERBERT RIVER RING-TAIL POSSUM
Pseudocheirus herbertensis

The Herbert River Ring-tail Possum is the emblem of the Queensland National Parks and Wildlife Service. This delightful creature is one of the three ring-tail marsupials belonging to the family Petauridae that are confined to upland tropical rainforests in north-eastern Queensland. All are nocturnal and predominantly vegetarian, feeding on leaves, blossoms and fruits. This species descends no lower than about 350 metres above sea level and has been recorded from the Seaview Range west of Ingham to Thornton Peak north of the Daintree River. Herbert River Ring-tails are typically all black or chocolate brown above and white below with a black head and white tip (below). In 1948, however, an American expedition discovered a uniformly caramel coloured form on ranges north of the Atherton Tableland; which was not recorded again until Stan and Kay Breeden photographed it in 1967.

During the day these lovely possums sleep in hollow trees or amongst clumps of vegetation including epiphytic ferns, and palm crowns. During early evening they emerge and after an intensive grooming session start out on their night-time feeding forays. Females generally raise two young at a time. Once the young are too large for their mother's pouch they cling to her fur or follow closely on her heels. Young possums become independant when they are not quite half grown.

Photo   :   Captive, Townsville.

GREEN RING-TAIL POSSUM
Pseudocheirus archeri

The Green Ring-tail Possum (opposite) ranges from Paluma, just north of Townsville, to the Mount Windsor Tableland north-west of Mossman and is rarely seen lower than 300 metres above sea level. The greenish colouration of this possum is due to an unusual combination of black, white and yellow pigments and the structure of its hairs; and so it differs from the Tree Sloth of South America, whose greenish fur is due to algal growth.

This possum is extremely difficult to spot as its green fur blotched with various shades of creamy white to yellow and broken up with stripes down the back, provides excellent camouflage for it within its leafy canopy home, particularly during the day when it sleeps curled up on a branch rather than in a den like some of the other possums. Its large protuberant eyes are well adapted for nocturnal vision and are easily picked up by spotlighting at night, as they reflect red light from the back of the eye. Green ring-tails generally produce only one young at a time which spends much time riding around on its mother's back in 'piggy-back' fashion after leaving her pouch. As the young grows its mother's fur becomes more and more tattered, as the fur is easily pulled out by the clutching youngster.

The third ring-tail species is the Lemuroid Ring-tail Possum, **Hemibelidus lemuroides**, which has been recorded from the Herbert River north to the Mount Spurgeon area west of Mossman at altitudes no lower than about 450 metres above sea level.

Photo   :   Mount Lewis

COPPERY BRUSHTAIL POSSUM
Trichosurus vulpecula

The Coppery Brushtail Possum (opposite) found in rth Queensland rainforests is a variety of the well known mmon Brushtail Possum which is widespread in the en eucalypt woodlands of Australia. One of the best alities to spot the Coppery Brushtail is at Mount pipamee National Park at 'The Crater', on the Atherton bleland. These appealing possums, easily recognised by ir prominent pointed ears, visit picnic tables at dusk ere they readily accept tasty titbits offered to them by itors. Their growls, grunts, squeals and a variety of other atter is often heard around this camping area. Their fur ies in colour from silver-grey to warm reddish brown. The ppery Brushtail, like the ring-tails, is mainly vegetarian d is known to eat leaves of Wild Tobacco which contain a ostance toxic to man but which obviously has no trimental effect on possums.

Coppery Brushtails belong to the same family of rsupials as the Grey and Spotted cuscuses, the alangeridae.

Photo   :   Mount Hypipamee National Park

GREY CUSCUS
Phalanger orientalis

The two species of cuscus, the Grey (below) and Spotted (page 64) are confined to rainforests of the Cape York Peninsula in Australia, and those of New Guinea. Early settlers thought cuscuses were monkeys when they first encountered them because of their large, almost forward-facing, eyes and small rounded ears. In fact, of course, there are no monkeys in Australia. During the day cuscuses are sluggish and sloth-like and spend most of their time sleeping amongst the dense vegetation with their tails coiled beneath them. At night they are surprisingly active, moving around the canopy and foraging on leaves and fruits. The occasional small bird or small mammal may also be included in their diet. When climbing their tails are uncoiled to aid balance and the bare scaly tail tip is well adapted for curling around and gripping branches for additional support, and is thus what is termed prehensile. Males and females of the Grey Cuscus are both uniformly greyish-brown and have a darker area of fur running down from their backs. Moreover, the skin is also greyish in this species.

Photo   :   Captive, Townsville.

SPOTTED CUSCUS
Phalanger maculatus

Males of this larger cuscus are easy to recognise as, unlike the plain grey females, their grey fur is blotched and spotted white. Unlike the Grey Cuscus the skin colour is a pinkish yellow and the ears are much smaller, barely protruding above the fur. These shy and secretive marsupials are more often heard clambering through the leafy tree tops in search of food similiar to those of the Grey Cuscus, than they are actually seen.

Photo   :   Captive, Townsville

CAPE YORK MELOMYS
Melomys capensis

Many people associate Australia with monotreme and marsupial mammals. Rats and bats which represent about 40% of our native mammal fauna are, however, true placental mammals (see above). To date 61 species of rodents have been described in Australia, all but 8 of which were probably here before any Europeans arrived.

Melomys rodents (left), of which there are 4 species in Australia, are predominantly tropical animals living in rainforests, woodlands, and grasslands. One species, the Thornton Peak Melomys, was discovered as recently as 1973, and is apparently confined to the upland rainforest of the Thornton Peak area between the Daintree and Bloomfield rivers. Forest melomys like the species pictured here, are predominantly tree climbers. Their long shiny hairless tail and broad hind feet assist them in their arboreal existence. Their typically rodent pair of incisor teeth efficiently gnaw through fruit, seed and occasional insect foods.

Photo   :   Iron Range

LARGE-EARED HORSESHOE-BAT
Rhinolophus philippinensis

Two species of **Rhinolophus** horseshoe bat are confined to areas of eastern coastal Australia; the most widespread of which is the Eastern Horseshoe-bat **Rhinolophus megaphyllus** which extends the whole length of the eastern coast from the tip of Cape York to south of the Victorian border. The illustrated species is found only from north of about Townsville. These bats are tiny-eyed insect eaters, flying at night in search of flying insects which they detect by using ultrasonic 'sonar' calls which bounce off objects and return signals to the most sensitive ears of these little 'flying mice'. The Large-eared Horseshoe-bat is found in various habitats, but is very much at home in tropical rainforests. It is commonly found roosting by day in relatively small colonies in caves, old mine shafts and even old buildings.

Photo   :   Paluma

FURTHER READING

ABERDEEN, J.E.C. 1979
An Introduction to the Mushrooms, Toadstools and Larger Fungi of Queensland.
Qld. Naturalist Club, Brisbane.

A.C.F. 1981 *Rainforest; Habitat.* A.C.F. Melbourne.

BARKER, J. & GRIGG, G. 1977
A Field Guide to Australian Frogs. Rigby, Sydney.

BREEDEN, S. & K. 1982
A Natural History of Australia : 1, Tropical Queensland. Second printing, Collins, Sydney.

COGGER, H.G. 1983
Reptiles and Amphibians of Australia. Third edition, Reed, Sydney.

COMMON. I.F.B. & WATERHOUSE, D.F. 1981
Butterflies of Australia. Revised Edition, Angus & Robertson, Sydney.

C.S.I.R.O. 1970
The Insects of Australia. Melbourne University Press, Melbourne.

FRANCIS, W.D. 1981
Australian Rainforest Trees.
Second Edition, Australian Government Printing Service, Canberra.

FRITH, D. & C. 1985
A Walk in the Rainforest. Tropical Australia Graphics, Paluma.

FRITH, C. & D. 1985
Australian Tropical Birds. Tropical Australia Graphics, Paluma.

HALL, L.S. & RICHARDS, C.C. 1979
Bats of Eastern Australia.
Queensland Museum Booklet No. 12, Queensland Museum, Brisbane.

HINTON, B. & B. 1980
A Wilderness in Bloom. Hinton, South Johnstone.

JONES, D.L. & CLEMESHA, S.C. 1981
Australian Ferns and Fern Allies. Second Edition, Reed, Sydney.

JONES, D.L. & GRAY, B. 1977
Australian Climbing Plants. Reed, Sydney.

PIZZEY, G. 1980
A Field Guide to the Birds of Australia. Collins, Sydney.

RUSSELL, R. 1980
Spotlight on Possums. University of Queensland Press, Brisbane.

RUSSELL, R. 1985
Daintree — Where the Rainforest meets the Reef.
Kevin Weldon & A.C.F., Sydney.

STRAHAN, R. 1983
The Australian Museum Complete Book of Australian Mammals.
Angus & Robertson, Sydney.

TRACEY, J.G. 1982
The Vegetation of the Humid Tropical Region of North Queensland.
C.S.I.R.O., Melbourne.

WILLIAMS, K.A.W. 1979, 1984
Native Plants of Queensland. Volumes 1 & 2, Williams, Brisbane.

Upland rainforest, Paluma

GENERAL INDEX TO PHOTOGRAPHS

Most famous of all North Queensland strangler figs is the 'Curtain Fig' which can be seen on the Atherton Tableland Here (opposite), the fig has developed on a sloping forest tree trunk and instead of sending all roots down the host's trunk many roots hang vertically downwards to the forest floor giving a 'curtain' effect.

CURTAIN FIG TREE Ficus virens
Stages in the development of the Curtain Fig Tree

PRESENT DISTRIBUTION OF AUSTRALIAN TROPICAL RAINFOREST

The map opposite indicates the approximate areas of tropical Australian rainforest remaining, which are in fact now le than half of that existing on the continent before the arrival of white man. Lowland rainforest, in particular, has been ve largely destroyed for agriculture and residential development. Less than half a percent of the entire Australian contine remains under tropical rainforest.

Place names given on the map are those of most mentioned in the text, or are those of significance from which tropic rainforest can easily be visited. Two most frequently visited tropical North Queensland centres are Townsville and Cair From the former; Paluma, 19 kilometres off the Bruce Highway at a point 60 kilometres north of Townsville, is the m convenient; and from Cairns fine lowland rainforest occurs immediately inland of the city and upland tropical rainforest is the Atherton Tableland. Many other rainforest locations are accessible from the coast between Ingham and Cairns and fro Cairns to Cape Tribulation. The McIlwraith ranges and Iron Range are truly magnificent forests more similar to those of N Guinea than Australia, but are accessible at present only to visitors by air, safari tours, or adventurous 4-wheel drive travelle

For information concerning readily accessible tropical rainforest locations, some with visitors' facilities, contact t Queensland National Parks and Wildlife Service offices at Brisbane, Townsville and Cairns.

HEADQUARTERS

National Parks and Wildlife Service,

P.O. Box 190,
BRISBANE. NORTH QUAY, Q. 4000

Brisbane   (07) 221 6111, 221 6236, 221 6465, 221 6629.

NORTHERN REGIONAL CENTRE

National Parks and Wildlife Service,

Pallarenda,
TOWNSVILLE, Q. 4810

Townsville   (077) 74 1332

CAIRNS REGIONAL OFFICE

Queensland National Parks and Wildlife Service,

P.O. Box 2066,
CAIRNS, Q. 4870

Cairns   (070) 53 4533